Iván Manrrique Hughes et al.

Sustainable BioRestoration

AF180329

Iván Manrrique Hughes et al.

Sustainable BioRestoration

Innovations in the concrete industry

ScienciaScripts

Imprint

Cover image: www.ingimage.com

This book is a translation from the original published under ISBN 978-3-639-53634-8.

Publisher:
Sciencia Scripts
is a trademark of
Dodo Books Indian Ocean Ltd. and OmniScriptum S.R.L publishing group

120 High Road, East Finchley, London, N2 9ED, United Kingdom
Str. Armeneasca 28/1, office 1, Chisinau MD-2012, Republic of Moldova, Europe
Managing Directors: Ieva Konstantinova, Victoria Ursu
info@omniscriptum.com

Printed at: see last page
ISBN: 978-620-6-76418-2

Sustainable BioRestoration

Innovations in the concrete industry

Iván Manrrique Hughes, Manuela Maldonado Torales, Karina Fernanda Crespo Andrada

Anabela Guilarducci, María. Gabriela Paraje (Editors)

CONTENTS

Summary

The study of climate change has highlighted the need to mitigate greenhouse gas emissions, especially in industries such as cement and concrete, responsible for 7% of total human-generated emissions. The search for sustainable alternatives has led to the exploration of biomineralization and the use of non-pathogenic microorganisms as "bio-restoration agents" in concrete structures. This process, in which microorganisms synthesize minerals, is classified into induced, controlled and biologically influenced mineralization. A multidisciplinary approach is crucial to overcome the challenges and advance sustainable conservation.

Key words:

- Climate change
- Greenhouse gases
- Cement industry
- Concrete
- Biomineralization
- Non-pathogenic microorganisms
- Sustainability
- Conservation
- Technological innovation

Chapter 1
The greenhouse gas emission problem

The study of climate change is a continuous process that has evolved over the last centuries marked by important discoveries and advances in the understanding of climate change. From the earliest indications of the relationship between global temperatures and the atmosphere, to the complex climate prediction models that exist today, research has played a fundamental role in the development of knowledge.

One of the initial discoveries that initiated these studies was the work of the French mathematician and physicist Joseph Fourier, who in 1824 published "Remarques générales sur les Températures du Globe terrestre et des espaces planétaires" (Fleming, 1999). In it he made one of the first definitions of the greenhouse effect, which consists of the atmospheric retention of the sun's heat, which leads to an increase in temperature in certain places. In 1861 John Tyndall presented his research "On the Absorption and Radiation of Heat by Gases and Vapours, and on the Physical Connexion of Radiation, Absorption, and Conduction" where it was experimentally demonstrated how certain gases, such as carbon dioxide (CO_2) and water vapor, could absorb and emit thermal radiation, which led to a deepening of the phenomenon explained by Fourier (Gentry, 1997). In 1896, Svante Arrhenius published his paper "On the Influence of Carbonic Acid in the Air upon the Temperature of the Ground", where he found that variation in the concentration of atmospheric CO_2 could cause large climatic changes in the global system (Anderson et al, 2016).

During the 20th century, the relationship between climatic evolution and human life began to be studied and an example of this is the work known as "The artificial production of carbon dioxide and it's influence on temperature" by the engineer Guy Stewart Callendar in 1938, which began to focus on the effects of anthropogenic activity on atmospheric systems and later, in 1961, Charles David Keeling developed "The Keeling curve", which is a reliable example of the constant increase of CO_2 due to human activity (Anderson et al., 2016).

In 1988, the Intergovernmental Panel on Climate Change (IPCC) began its work, which was born from two United Nations agencies, the World Meteorological Organization (WMO) and the United Nations Environment Programme (UNEP). The IPCC's objective is to generate knowledge based on scientific evidence in order to provide quality information to decision makers and its work is condensed annually in its "Assessment Reports" which detail the current situation and future predictions (Agrawala, 1998).

Currently, climate change is defined as a complex phenomenon that occurs as a result of the emission of greenhouse gases into the atmosphere, mainly through human activities such as the burning of fossil fuels, deforestation, agriculture, industrialization, among others, with CO_2, methane (CH_4) and nitrous oxide (N_2O) being the gases that have the greatest impact, as they possess properties of absorbing long wave infrared radiation emitted by planet Earth, resulting in the retention of heat within the atmosphere, exerting an effect similar to that of an insulating layer that limits heat loss to outer space (Yoro, 2020). In addition, the global warming triggers feedback processes that lead, for example, to the development of *permafrost* melting, which releases large amounts of CH_4, which has a much more potent effect than CO_2. The release of CH_4, from melting *permafrost* in regions such as Siberia and Alaska, represents a significant threat to global climate stability (Schuur et al., 2015). CH_4 is a greenhouse gas, 28 to 34 times more effective at trapping heat than CO_2 over a 100-year period. In the latest climate records, temperatures were observed to exceed 3 °C in the last 40 years, leading to the thawing of vast tracts of *permafrost* that have remained frozen for millennia (IPCC, 2021).

Chapter 2

Environmental aspects related to concrete

The growing increase in the demand for infrastructure and housing in large urban centers leads to the assumption that these quantities of ready-mixed concrete are projected to increase. As an example, one can mention a forecast indicating a global concrete production of 18 billion tons by the end of 2050 (Wang et al., 2017).

This massive consumption of materials such as concrete presents the main problem of environmental impact. The concrete industry consumes a large amount of natural materials, resources, energy and even capital; it generates large amounts of gaseous emissions and waste from the repair and demolition of structures when they are not properly managed, which entails significant social costs and environmental burdens. In this regard, it should be noted that approximately 11% of greenhouse gases come from this industry and that more than 5% of the global anthropogenic carbon dioxide (CO_2(g)) generated annually corresponds to the production of Concrete (Van Belleghem et al., 2017; Córdoba et al., 2023).

Of all the processes associated with concrete production, Portland cement manufacture has the dominant environmental impact (especially in terms of $CO_{2(g}$ emissions). It is estimated that energy consumption in the production of 1 t of cement is 3.1 to 5 GJ and that it generates between 0.73 and 0.99 t of $CO_{2(g)}$ emissions due to the decomposition of limestone during production and the high temperatures required in the kiln (close to 1450°C). Thus, the cement industry accounts for approximately 10% of current anthropogenic $CO_{2(g)}$ emissions and 12 to 15% of total industrial energy.(Wang et al., 2017; Li et al., 2019).

It is therefore recognized that the construction industry is responsible for significant environmental impacts, so it is important that alternatives are explored to reduce these effects. For the cement sector, this includes cleaner production, recycling, and lower impact cements (Wang et al., 2017. In this context, the partial replacement of Portland cement clinker with mineral additions is a strategy to reduce the environmental impact of the construction industry. This strategy has the advantages of reducing energy consumption and increasing production without the need for new kilns (Scrivener et al.; 2018).

Chapter 3

Cement industry: process description, energy consumption and global warming implications

During the 21st century the effects of climate change have become more severe having a great impact on both the environment and its relationship with society (IPCC, 2019). This led to an increased need to reduce and mitigate GHG emissions considering the negative consequences this implies on life on earth. The Environmental Protection Agency (EPA) indicated that during the year 2021 the industrial sector was in charge of more than 20% of total GHG emissions and discretizing only for the cement industry it is found that due to its high energy consumption the same is responsible for 7% of such emissions. Taking 2018 as a reference, it is obtained that an approximate of 1.50 Gt CO_2eq were due to the cement industry in its contribution to concrete production processes (Andrew, 2019).

Each stage of the cement production process contributes to global warming both directly and indirectly (Tayebani et al, 2023). This is why it is necessary to make a general overview of the industrial process carried out to obtain this material. Cement production plants are often strategically located near quarries that offer an abundance of limestone rocks and clays, essential raw materials for the production process. This choice of location is critical because it leads to a significant reduction in transportation costs by minimizing the distance that materials must travel to the production plant. Such a reduction not only has a direct impact on operating expenses, but also contributes to the mitigation of the carbon footprint related to the transportation of large volumes of materials. In addition, proximity to raw material sources ensures a stable and reliable supply, a determining factor for the continuity and security of the supply chain. It is also essential to consider local regulations and to undergo an Environmental Impact Assessment, as required by the competent authority, to ensure compliance with environmental standards and the viability of the site chosen for the development of this type of industry.

The extraction of these materials is carried out using mining techniques that vary according to the geological and topographical characteristics of the deposit in question. Some techniques used in this step can be open pit mining, which is commonly used in limestone deposits, as well as subway mining, which may be necessary in areas where limestone is found at greater depths (Tayebani et al, 2023). Some machinery used in this phase are usually excavators, dump trucks and drills to remove the material and transport it to processing facilities. If the quarry and the plant are adjacent to each other, a conveyor belt is usually installed to move the material, and if the quarry is a little further away, transport by truck may be used (Griffiths et al, 2023). Once the raw material is transported to the production plant, it undergoes a primary crushing process, where its size is generally reduced by gyratory crushers. This allows greater efficiency for the next step, known as grinding, where the crushed material is fed into ball or roller mills for fine grinding. This process homogenizes the mixture and increases the specific surface area of the raw material to facilitate the chemical reactions that will be carried out by the thermal processes of the next step (Mohamad et al, 2022).

One of the fundamental steps in the production of cement is the use of rotary kilns where the raw material is introduced with the aforementioned treatment. The mixture is introduced into a cylindrical kiln that is inclined and rotates slowly on its longitudinal axis. It enters through the upper part. Inside the kiln, which reaches temperatures of around 1450 °C, the clinkering process takes place. This process consists of a series of chemical reactions between the components of the raw material that lead to the formation of new compounds, such as silicates, aluminates and calcium ferrites, which are responsible for the setting of the cement (Bojanovský et al, 2022). The chemical processes involved in this stage are as follows:

Decarbonation:

The calcium carbonate ($CaCO_3$) present in limestone decomposes thermally at high temperatures, releasing CO_2 and leaving calcium oxide (CaO) as a product:

$$CaCO_3 \rightarrow CaO + CO_2$$

Silicate and Aluminate Formation Reactions:

CaO, silica (SiO_2), alumina (Al_2O_3) and iron oxide (Fe_2O_3) react to form calcium silicates and aluminates and other compounds. These reactions are complex and can occur in several stages, but in general, they involve the combination of these components to form different crystalline phases, such as:

Tricalcium silicate (C_3S):

3 CaO.

Dicalcium silicates (C_2S):

$2\ CaO.SiO_2$

Tricalcium aluminates (C_3A):

Al_2O_3. 3 CaO

Tetracalcium ferrites (C_4AF):

$Al2O_{(3)}$-4 CaO-Fe_22O_3

The composition of clinker in cement is a fundamental aspect in the construction industry, where aluminates and silicates define the properties and performance of the material. C_3A is known for its fast hydration rate and high heat released during the process making it responsible for the initial setting of the cementitious paste. However, its vulnerability to sulfate attack can compromise the durability of cement-based structures. As for C_4AF, it has a high chemical stability and its presence in the clinker composition depends on the proportion of aluminum and iron oxides, although its contribution to strength development is lower than that of silicates (Sun et al, 2023).

As for silicates, C_3S has high hydraulic activity and high reaction rate, conferring short- and long-term strength and affecting the onset of setting. On the other hand, C_2S has slower hydration and is the main component of slow-setting cements, contributing significantly to long-term strength and durability (Laanaiya et al, 2019).

Reduction and Oxidation Reactions:

In addition to compound formation reactions, reduction and oxidation reactions can occur that influence the composition and properties of the clinker.

Reduction Reactions:

$$Fe_2O_3 + 3\,CO \rightarrow 2\,Fe + 3\,CO_2$$

Oxidation Reactions:

$$C + O_2 \rightarrow CO_2$$

The resulting clinker is a solid mass of grayish colored granules that is rapidly cooled in a rotary cooler to avoid the formation of undesirable phases (Griffiths et al, 2023). These phases encompass various problems, such as overheating phases, which occur when the temperature in the kiln is excessively high or when the residence time of the material is prolonged, which can result in agglomeration of the clinker or the formation of unwanted crystals. Also, incomplete reaction phases represent a challenge because, if the clinkerization conditions are not suitable, some chemical reactions may remain uncompleted, resulting in the presence of non-reactive phases or partially formed in the clinker, which affects its strength and durability. Another detrimental process can be a contamination in the mixture, generating unwanted phases, where impurities in the raw material or the introduction of contaminants can compromise the quality of the clinker and cement. In addition, improper crystallization phases, generated by inadequate cooling of the clinker, can cause an undesirable microstructure and negatively affect its mechanical and chemical properties. Therefore, it is essential to constantly monitor clinkerization conditions to minimize the formation of these undesirable phases, thus ensuring the quality and consistency of the final clinker and cement (Bisulandu & Huchet, 2023).

Finally, the cooled clinker is stored in silos before grinding, to which a small amount of gypsum ($CaSO_4$) is added to regulate the setting time of the final product known as cement. The proportion of aggregate ranges between 3% and 5% according to the final product to be achieved. This process repeats the use of ball or roller mills. This will be packaged in bags or transported in bulk for distribution according to the consumer's needs (Tayebani et al, 2023).

The specific energy consumption in cement production varies from one technology to another. The dry process uses more electrical energy but much less thermal energy than the wet process. In industrialized countries, primary energy consumption in a typical cement plant is up to 75% fossil fuel and up to 25% electrical energy using a dry process. Pyrolytic processing consumes most of the thermal energy used, which is equivalent to approximately 91.6% of the total fuel consumption involved in this phase. However, electrical energy is used mainly to operate both the raw material crushing and grinding equipment 33% and the clinker equipment 38%. As a consequence, significant amounts of GHGs are released into the atmosphere due to the burning of fossil fuels to supply the energy needed by the cement industries. The specific thermal energy consumption in cement industries is around 4 and 5 GJ/tonne of fuel supply to maintain the pyrolytic process (Madlool et al, 2011).

Reviewing the total energy consumption with respect to each stage, it is observed that 1.4% corresponds to the extraction of materials in quarries, 2.8% is related to the crushing and grinding processes, 91.6% is attributed to the use of the pyrolytic kiln, 3.2% is allocated to the clinker grinding process, and the remaining 1% is used in the storage and distribution of the final product. This distribution highlights the significant energy demand associated mainly with the operation of the pyrolytic kiln, underlining the importance of improving efficiency in this aspect to achieve greater sustainability in cement production (Afkhami et al, 2015).

Cement companies are making efforts to make their production more sustainable due to the economic, environmental and social benefits that this entails. However, often, the implementation of these modifications implies a great expense for the producers, which leads them to be discarded due to their low profitability. In spite of this, some viable alternatives have been developed such as the use of alternative fuels in the manufacturing process (Kahawalage et al, 2023). These materials have to possess pozzolanic (siliceous or alumino-siliceous materials) or latent hydraulic properties (Gartner & Sui, 2017). The use of these materials contributes to the mechanical properties of cement and they react with calcium hydroxide ($Ca(OH)_2$) released during hydration to form additional cementitious compounds. This dual functionality improves the strength and durability of concrete, one of the main materials that uses cement in its composition.

Chapter 4

Concrete as a main cementitious base product: structural problems and challenges to be addressed

The growth of the world's population and the development of new infrastructure related to the construction industry will generate increased demand for cement. It is therefore crucial to seek a more sustainable market and develop new solutions to address the sector's challenges.

Concrete is one of the most widely used cement-based building products due to its accessibility, affordability, and superior mechanical, thermal, and insulation properties compared to other materials such as steel and wood (Sohail et al., 2018).

Compared to steel, concrete has the possibility of being produced locally as its inputs are often widely available in many places and this has consolidated it as the most used material in the world after water, with a production magnitude of approximately 10 billion tons per year worldwide. Concrete, composed mainly of cement, aggregates (such as gravel and sand) and water, is one of the most versatile and widely used building materials in the world. One of the key characteristics of concrete is its setting time, which marks the beginning of the material's hardening and strength acquisition. Concrete setting is a complex chemical process that occurs when water combines with cement, triggering hydration reactions that form chemical bonds and crystalline structures. As these reactions progress, the concrete changes from a plastic to a solid state, allowing it to be handled and molded during the construction process. The setting time varies with concrete composition, environmental conditions and other factors, and is crucial in determining the time available for transporting, pouring, molding and finishing the concrete before it begins to harden (Amran et al, 2021).

This is the construction material of choice because it is versatile and offers a variety of formulations adapted to different needs and conditions. First, conventional concrete, composed of cement, aggregates and water, is widely used in various construction applications. On the other hand, high-strength concrete is characterized by higher compression, ideal for structures that require additional strength, such as bridges and tall buildings. In addition, low permeability concrete, with its reduced capacity to absorb water, is suitable for wet or corrosive environments, such as tunnels and swimming pools. Self-compacting concrete, which flows and fills the forms without the need for vibration, is ideal for smooth and uniform finishes, especially in architectural elements. On the other hand, fiber-reinforced concrete improves tensile strength and durability, making it ideal for industrial pavements and containment structures. Finally, lightweight concrete, with reduced density thanks to lightweight aggregates, is an option for elements with reduced structural loads, such as ceilings and precast elements. Each type of concrete presents specific characteristics that make it suitable for different construction applications, providing versatile and efficient solutions for various needs in the construction industry (Sohail et al, 2018).

Concrete is characterized by its compressive strength but weak tensile strength. Because of this, it is prone to cracking, affecting the overall strength of the structure and its service life. Cracks can form both internally and externally due to plastic shrinkage, thermal stresses, drying settlement, weathering, etc. The problem with cracking of a structure lies in the fact that its capacity to absorb traction decreases and increases with time. Once these cracks are accentuated, both liquids and gases can enter aggressive substances into the concrete matrix producing a reduction of the mechanical resistance leading to a possible collapse. Crack repairs in materials can be costly and can occur in locations that are difficult to inspect and reach (Triantafyllou et al, 2017).

Cracks in concrete, both surface and internal, represent a significant engineering challenge due to their potential consequences in terms of durability, strength and safety of structures. According to a report by the Federal Highway Administration, the United States spends $4 billion annually in direct maintenance costs for concrete highway bridges. According to De Rooij and Van Tittelboom, the United Kingdom spends 45% of its annual construction cost on maintenance of existing concrete structures (de Rooij et al., 2013; Zhang et al., 2020).

When we talk about surface cracks, their most common reasons may be shrinkage cracking, a phenomenon that occurs during the setting and curing process of concrete when the water present in the mix evaporates, causing the material to shrink. This shrinkage can generate surface cracks, especially if it is not uniform or is restricted by reinforcement or other adjacent structures (Kayondo et al, 2019). Another reason refers to external overloads or impacts, such as heavy vehicular traffic, falling heavy objects or seismic forces as these may exceed the design load capacity of the concrete (Meng et al, 2019). Also, there is the possibility of cracks being generated from freeze-thaw cycles as the water present in the pores of the material can repeatedly expand and contract thus generating pressure cracks (Wang et al, 2022).

Internal cracks can be referred to as internal shrinkage cracking, which is a phenomenon that occurs during the setting and curing of concrete and can result in microscopic cracks that, although initially imperceptible, can propagate over time due to applied loads or structural movements. Another reason may be caused by alkali-aggregate reactions

(AAR), where certain aggregates react with the alkalis present in the cement, forming expansive gels that can cause internal cracks in the material. Finally, it is also possible when design deficiencies in the construction of the structure lead to concentrations of internal stresses that eventually manifest themselves as cracks in the concrete (Askar et al, 2023).

Chapter 5

Innovative and sustainable searches for the construction industry: biologically induced calcium carbonate mineralization as an alternative

Traditional methods of solving this problem often raise significant concerns because they involve the use of synthetic materials and chemicals that can cause environmental damage. For example, the use of epoxies and other chemical sealants can release volatile organic compounds (VOCs) that contribute to air pollution and can be harmful to human, animal and environmental health. Another problem is that some repair methods may require the removal and replacement of entire sections of concrete, which generates large amounts of waste and consumes a significant amount of energy and raw materials (Alemu et al, 2022).

Currently, in the search for sustainability in the construction industry, several strategies are under study: using less cement or looking for substitutes, adopting carbon capture techniques, using recycled materials, exploring nanotechnology and incorporating biotechnology (Ahmed et al, 2021). These practices are not only aligned with the environmental objectives of ONE HEALTH, but also respond to the growing demand for responsible and resource-efficient solutions, so researchers from different areas recognize their importance for the economic benefits they provide, the reduction of long-term environmental impacts and their contribution to a more sustainable built environment (Adnyana et al., 2023).

The process known as biomineralization is the synthesis of certain minerals (more than sixty different biominerals) with very specific functions both in animals, as those capable of forming a greater number of different biominerals, followed by bacteria, vascular plants and, finally, fungi and protozoa with calcium being the major cation, followed by iron and according to the anion, it is phosphates, followed by oxides and carbonates. Generally speaking, $CaCO_3$ polymorphs, such as aragonite and calcite, are widely distributed in almost all Phyla. In contrast, halides are rare and only observed in a few groups, while sulfides and some oxides have been identified exclusively in prokaryotes (Handore et al, 2021).

This process can be classified as follows:

1) Biologically Induced Mineralization (BIM), where minerals are formed extracellularly as a result of the metabolic activity of the OM.

2) Biologically Controlled Mineralization (BCM), minerals are formed within organic matrices and the MO can regulate extra-, inter- or intracellularly the growth of minerals.

3) Biologically mediated mineralization (BMM), where the presence of organic matter on the cell surface influences the formation, morphology and composition of the mineral.

In the first group is the use of non-pathogenic microorganisms (MO) as "restorative agents" in cement-based composites as a promising alternative. BIM is widespread among the various Phyla of the Bacteria domain, both in oxic, anoxic and interphase oxic-anoxic environments. It also occurs in some eukaryotes, fungi and algae, and animals, but in the latter it is less frequent compared to prokaryotes. Specifically, the ability to induce microbiological precipitation of $CaCO_3$, known as MICP (Microbially Induced Calcium Carbonate Precipitation) from different MOs, presents an innovative potential to improve the durability and strength of concrete and reduce the dependence on synthetic materials and harmful chemicals, minimizing their impacts (Zhang et al, 2023).

The discovery of the MICP technique dates back to the 1990s when a study group in Paris sought to demonstrate that bacteria were capable of precipitating calcite on damaged limestone surfaces. At the beginning, the aim was to use this technique in the restoration of cultural heritage (Ortega-Villamagua et al, 2022). During the following years it was demonstrated that there is a great variety of bacterial genera and species capable of being used in this field of application such as *Pseudomonas, Bacillus, Vibrio*, *Streptomyces*, *Pantoea, Cupriavidus, Myxobacteria* and *Halobacillus, among* others (Soffritti et al, 2019).

MICP is a complex process in which a wide variety of MO can participate, triggering a series of biochemical and metabolic pathways and can theoretically occur both by BCM and by BIM. But most of the literature focuses on the latter since the formation of the mineral occurs extracellularly, the most commonly used being ureolytic bacteria, since they have the ability to induce $CaCO_3$ precipitation. These bacteria use the enzyme urease to catalyze the hydrolysis of urea (CON_2H_4) into CO_2 and ammonia (NH_3), resulting in an increase in pH in the medium and subsequent precipitation of CaCO3. This process, known as heterogeneous nucleation, occurs when excreted $CaCO_3$ acts as a crystallization nucleus on walls, bacterial membranes or cell debris. This happens when a level of supersaturation is reached, where the amount of available calcium and carbonate ions exceeds the dissolving capacity of the water, leading to supersaturation of the system. As the ion concentration increases it facilitates the formation of crystals in the cell wall thus inducing mineralization (Gebru et al, 2021) (Fig. 1). MICP can be classified as passive or active, in the first case occurring in a non-specific binding of cations and recruitment of anions into solution, resulting in nucleation and growth of the mineral on the surface. Active mineralization is dependent on the redox activity of metal ions resulting from metabolic activities that form minerals on bacterial surfaces.

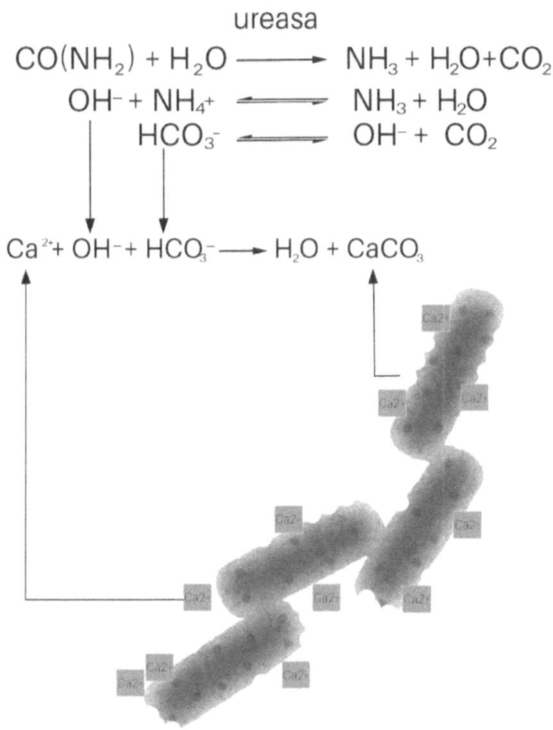

Fig.1: Biologically induced $CaCO_3$ mineralization process.

The $CaCO_3$ precipitation process is regulated by a complex interaction between chemical and biological factors. The main chemical factors include calcium ion concentration, pH and the availability of nucleation sites. On the other hand, biological factors, such as the type of MO, nutrients and the concentration of dissolved inorganic carbon (DIC) produced by the bacteria, are crucial for their survival in hostile environments such as that provided by cement (Gebru et al, 2021). Calcium plays a key role in influencing the concentration of carbonate ions to generate the saturation state of the system, which can come from the environment or from added nutrients. CID concentration is influenced by temperature and CO_2 partial pressure, as it forms complexes with calcium ions, reducing $CaCO_3$ saturation and enhancing calcite precipitation. The pH is a determinant in the biochemical activity and survival of MOs. It is also critical as $CaCO_3$ tends to precipitate at high pH levels rather than dissolve because microbial metabolism increases alkalinity, which shifts the equilibrium toward crystal precipitation (Muller et al, 2022). MOs can be nucleation sites *per ser*, due to the negative charge on their surface, which allows the accumulation of calcium and carbonate ions on their membranes and cell walls, promoting the formation of calcium carbonate therein. In Gram-positive bacteria, the negative charge comes from the carboxyl or phosphate groups on the teichoic and lipoteichoic acids, while in Gram-negative bacteria, this charge is due to the phospholipids and lipopolysaccharides present in the membrane. The strong electrostatic affinity to attract divalent cations, allows the accumulation of calcium ions on the cell wall surface, achieving a state of supersaturation of calcium ions which then binds to carbonate ions resulting in the formation of $CaCO_3$ in the cell wall, however, they can also be controlled from the formulation of mortars and concretes (Muller et al., 2022).

As for MOs, they can be classified into different groups according to their metabolic mechanisms. Photosynthetic ones, associated with cyanobacteria, which remove CO_2 from the water layer adjacent to the cell, altering the microenvironment by increasing the pH. This causes an imbalance of carbonic acid (H_2CO_3), leading to the production of $CaCO_3$ in the presence of calcium ions. MOs, such as *Bacillus pseudofirmus* and *Bacillus cohnii*, employ organic acids in the metabolic conversion of organic compounds (organic acid salt) into $CaCO_3$ by aerobic oxidation with CO_2 generation, resulting in $CaCO_3$ production in an alkaline environment (Addenan et al., 2023). Sulfate-reducing bacteria, responsible for sulfate reduction as a terminal electron acceptor, are involved in the mineralization of organic matter under anaerobic conditions (Paul et al. 2017). Finally, bacteria involved in the nitrogen cycle, especially ureolytic bacteria are particularly promising for carbonate precipitation, using urea hydrolysis as the main metabolic pathway (Naveed et al., 2020).

The incorporation of nutrients, such as carbon, nitrogen or calcium sources, also influences microorganism-induced precipitation. Together, these factors contribute to the self-repairing capacity of the materials and are closely linked to long-term viability in environmental conditions unfavorable for microbial growth.

As can be seen from the above, MICP is a process that is highly influenced by the interaction between the OM and its environment, so it is extremely important to understand the different biochemical and metabolic pathways involved in the process, the growth kinetics, as well as the nutritional and environmental requirements, and the survival mechanisms of ureolytic bacteria.

The genus *Bacillus* is characterized by having this property of forming endospores that allows it to survive in harsh environments. This ability is beneficial in this context, as the endospores can withstand extreme conditions such as high temperatures, lack of nutrients and exposure to chemicals (Akindahunsi et al, 2021). In addition to their resistance, this genus also has the ability to extract energy from the oxidation of organic compounds with their main metabolic pathways being calcium lactate $Ca(C_6H_{10}O_6)$ and calcium acetate $Ca(C_4H_6O_4)$ (Muller et al, 2022). These metabolic pathways allow them to use organic compounds as carbon and energy sources for their growth and metabolism.

<u>Biochemical equations involved from the oxidation of $Ca(C_6H_{10}O_6)$ for the formation of $CaCO_3$:</u>

$$CaC_6H_{10}O_6 + 6\ O_2 \rightarrow CaCO_3 + 5\ CO_2 + 5\ H_2O$$

$$5\ CO_2 + 5\ Ca(OH)_2 \rightarrow 5\ CaCO_3 + 5\ H_2O$$

$$Ca(C_4H_6O_4) + 4\ O_2 \rightarrow CaCO_3 + 3\ CO_2 + 3\ H_2O$$

$$3\ CO_2 + 3\ Ca(OH)_2 \rightarrow 3\ CaCO_3 + 3\ H_2O$$

The first equation shows the oxidation of $Ca(C_6H_{10}O_6)$ in the presence of oxygen (O_2) to produce $CaCO_3$, CO_2 and water (H_2O). The second, as the CO_2 produced in the first equation reacts with $Ca(OH)_2$ present in typical concrete mixtures to form $CaCO_3$ and H_2O. The third represents the oxidation of $Ca(C_4H_6O_4)$ in the presence of O_2 to produce $CaCO_3$, CO_2 and H_2O and the last, how the produced CO_2 e reacts with

$Ca(OH)_2$ to form $CaCO_3$ and H_2O (Fig. 2). As can be seen, calcium-containing nutrients are necessary for the bacteria to perform MICP and can use calcium lactate, calcium nitrate, calcium chloride and calcium acetate, among others (Muller et al., 2022).

The $CaCO_3$ crystals produced by MICP can be divided into different polymorphs. These include calcite, vaterite and aragonite along with hydrated crystalline phases such as monohydro calcite and ikaite, as well as various amorphous calcium carbonate (ACC) phases which have significant differences in their thermodynamic stability and physical properties (Seifan & Berenjian, 2018). The formation of the aforementioned crystals can be influenced by the bacterial cell wall, its extracellular polysaccharide (EPS) content, the composition of the culture medium, bacterial metabolic activity, etc. The most recurrent case in MICP is the precipitation of vaterite although it can also under certain conditions evolve to calcite, ACC can also be found (Seifan & Berenjian, 2018).

It was observed that vaterite tends to form at low pH levels, while calcite is more prevalent at higher pH levels. The addition of different conditions can also influence crystal morphology, with the use of acid favoring calcite formation. Calcium chloride predominantly induces precipitation of vaterite (Wen et al, 2019).

Fig. 2: Crystal structure of $CaCO_3$.

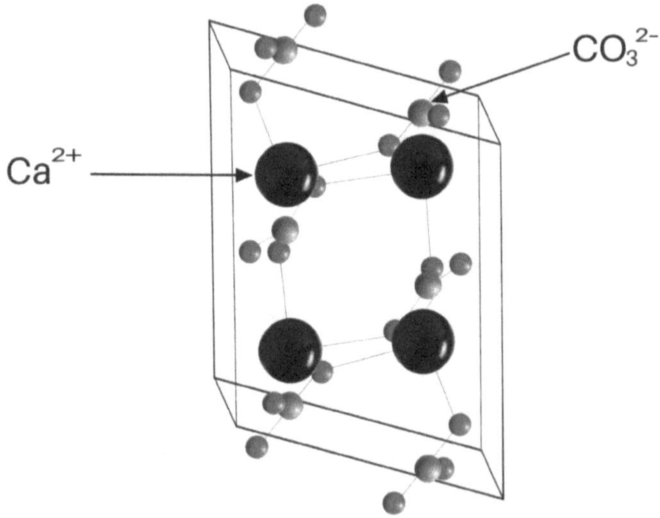

Chapter 6

Lysinibacillus sphaericus: a promising microorganism in microbiologically induced calcium carbonate precipitation.

Lysinibacillus sphaericus, formerly known as *Bacillus sphaericus*, underwent a genus change due to its distinctive composition of high lysine, alanine, aspartic acid and glutamic acid content in its cell wall peptidoglycan, as well as phylogenetic and physiological analyses. It is a Gram positive bacillus, sometimes observed as Gram variable, strict aerobic, non-pathogenic, low Guanine+Citokine, producing high levels of urease. A distinguishing feature is its ability to form spherical spores that are terminally located in deformed sporangia (Fig. 3). In addition, this bacillus has a limited ability to utilize various sugars, such as glucose, xylose, galactose, mannose, fructose, sucrose, lactose and sorbitol, as carbon and energy sources. *L. sphaericus* is catalase and oxidase positive, does not produce hydrogen sulfide (H_2S) or indole, and is negative for nitrate to nitrite reduction and for galactosidase enzyme activity. These characteristics contribute to define and distinguish this species within the genus *Lysinibacillus* (Hernandez-Santana et al., 2016; Allievi et al., 2014).

In the last decade, it has emerged as a MO of interest due to its remarkable versatility. Its ability to produce a variety of secondary metabolites with significant biological properties stands out. One of its most prominent uses is based on the application to produce biopesticides effective against mosquito larvae transmitting diseases such as dengue and malaria (Peña-Montenegro et al, 2015). It can also be applied in bioremediation of contaminated soils , plant growth promotion and pathogen control in agriculture (Nusrat et Shimizu, 2021).

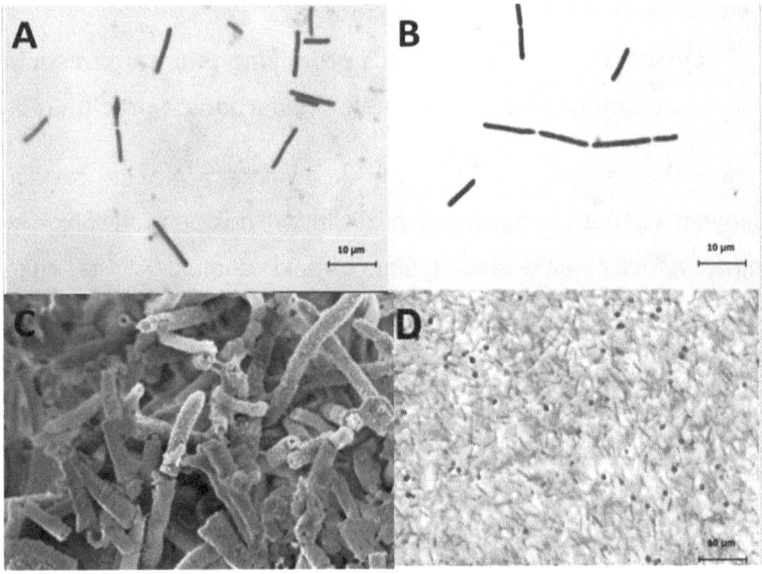

Fig. 3: A and B) Gram staining of *Lysinibacillus sphaericus* Magnification 1000x. C) Scanning Electron Microscopy (SEM) and D) Spores of *L. sphaericus* with Wirtz staining. Scale 10 µm. Magnification 1000x.

L. sphaericus is one of the most studied and used microbial species in MICP technology. Its ability to precipitate $CaCO_3$ offers great potential in the repair of cracks in cement-based structures, a phenomenon called biocementation (Zhang et al, 2023). By having a high efficiency in the formation of this mineral and having the ability to adapt to different environmental conditions, it offers an alternative to problems in the field of environmental-civil engineering. It is because of this that it can work efficiently in environments with pH fluctuations and high concentrations of metal ions making its use possible in different contexts of construction and rehabilitation of infrastructures (Marcondes et al, 2021).

Proper development of this new biotechnological tool can lead to long-term cost reductions for the industry by improving the durability of concrete and could even develop synergistic mixtures with admixtures currently used in the market. However, there are also important challenges and considerations to take into account. For example, precise control over conditions is necessary for proper precipitation and to provide reliable and reproducible results (Zhang et al, 2023). Other points to be taken into account for process improvement is to adapt the complexity of the process to industrial conditions in order to optimize the repair times that could be carried out once introduced into the cementitious matrix (Mokhtar et al, 2021). Considering that it is a biological material, its application should be sought in conditions that are not extreme for the development of its life cycle, since these can lead to the inactivity of the MO.

Another important point is to analyze the control of $CaCO_3$ distribution, since the production of acids in certain parts of the mixture can lead to the degradation of the cementitious material, just as an increase in pH near the steel of a reinforced concrete can favor the corrosion process of the same, generating a potential impact on its strength and durability (Green, 2020). The activity of biocementation will also depend on the ability of bacteria to move freely, either by injection through the pore space or by particle-particle contact, allowing self- repair of cracks from the inside, resulting in homogeneity of the treated materials, or from the outside, sealing fragmented surfaces exfoliated by environmental and anthropogenic agents. These conditions require a balanced relationship between microbial and pore size between sand particles producing a $CaCO_3$ layer that protects the deteriorated piece against water absorption and consolidate its internal structure, and can potentially be used to prevent or even reverse the deterioration of surfaces, bioconserving them (Xu, J. M., et al, 2014). However, the action of the MICP process on concrete at early ages is still an approach of great potential, since the microstructure of the material could be improved by decreasing porosity or in the self-repair of microcracks that arise in the setting and hardening process, increasing durability (Pise et al., 2021).

The establishment of and compliance with the parameters defined by standards and technical regulations (national and international) play a fundamental role in the concrete industry, particularly in applications such as MICP. These standards provide a regulatory framework that ensures quality and safety in the use of materials and processes in the construction of concrete structures. In the context of MICP, compliance with the limits set by the standards becomes a crucial challenge to ensure the viability and effectiveness of the concrete self-repair process. In this sense, it is important to establish the balance between the requirements to be met by the concrete component materials and the requirements to promote MICP (Zhang et al., 2023).

For example, keeping solid residue levels within allowable limits is a major issue to consider, as high concentrations can inhibit the bacterial activity necessary for $CaCO_3$ precipitation. This poses a conflict in terms of ensuring the quality of the mixing water without compromising the effectiveness of the MICP process. In addition, control of the pH of the mixing water is crucial to the success of MICP, but it must always be within the pH range allowed by regulations to be used as a concrete component. Maintaining the pH within the permitted range is a continuous challenge, as any deviation can significantly affect the ability of bacteria to induce $CaCO_3$ precipitation. Compliance with sulfate and chloride limits is also essential to avoid MICP inhibition while maintaining established minimums to ensure concrete durability. However, achieving these goals can be challenging due to the variability of water sources and the need to implement appropriate treatment techniques. With respect to iron and alkalis, their presence in the mixing water can adversely affect the effectiveness of the MICP process and be corrosive to the concrete. This poses an additional challenge in terms of carefully selecting the materials and admixtures used in concrete production to minimize the introduction of these elements (Wei et al., 2023).

As mentioned above, a new alternative repair biotechnology tool has been described in recent years with the use of non-pathogenic MOs as "bio-restoration agents" to improve surfaces that have been fragmented and exfoliated by their interaction with the aggressive physical, chemical and biological agents of the surrounding environment . MICP has great potential as a tool for the alternative restoration of concrete structures, but much information remains to be elucidated about the survival of MOs in the hostile environment provided by the cement, being necessary to deepen the study. For this reason, the study of MO and the optimal conditions of growth and survival is of great interest as a possible biotechnological tool for alternative repair aimed at improving the conservation and/or restoration of concrete structures of patrimonial, scientific and technical interest, being necessary a multidisciplinary approach to work in different areas of science and technology, integrating concepts of microbiology, environmental and civil engineering, chemistry and geology, to find innovative and sustainable solutions to the complex challenges that are currently faced (Iqbal et al., 2021).

Chapter 7
Bibliography

Addenan NA, Ngalimat MS, Rahman RNZRA, et al. Evaluation of Calcium Carbonate Precipitation by *Bacillus spp.* Isolated from Stingless Bee Products. Sains Malaysiana. 2023;52(6):1723-1735. doi:10.17576/jsm-2023-5206-09.

Adnyana IMdm, Utomo B, Eljatin DS, Sudaryati NLg. One Health approach and zoonotic diseases in Indonesia: Urgency of implementation and challenges. *Narra J. 2023;3(3).* doi:10.52225/narra.v3i3.257.

Afkhami B, Akbarian B, Beheshti A, Kakaee AH, Shabani B. Energy consumption assessment in a cement production plant. *Sustainable Energy Technologies And Assessments.* 2015;10:84-89. doi:10.1016/j.seta.2015.03.003.

Agrawala, S. Context and Early Origins of the Intergovernmental Panel on Climate Change. *Climatic Change.* 1998. *;* 39, 605-620. https://doi.org/10.1023/A:1005315532386.

Akindahunsi AA, Adeyemo SM, Adeoye A. The use of bacteria (*Bacillus subtilis*) in improving the mechanical properties of concrete. *Journal Of Building Pathology And Rehabilitation.* 2021;6(1). doi:10.1007/s41024-021-00112-7.

Alemu D, Demiss W, Korsa G. Bacterial Performance in Crack Healing and its Role in Creating Sustainable Construction. *International Journal Of Microbiology.* 2022;2022:1-10. doi:10.1155/2022/6907314.

Allievi MC, Palomino MM, Acosta MP, Lanati L, Ruzal SM, Sánchez-Rivas C. Contribution of S-Layer Proteins to the Mosquitocidal Activity of *Lysinibacillus sphaericus*. *PloS One*. 2014;9(10):e111114. doi:10.1371/journal.pone.0111114.

Amran M, Murali G, Khalid NHA, et al. Slag uses in making an ecofriendly and sustainable concrete: A review. *Construction & Building Materials*. 2021;272:121942. doi:10.1016/j.conbuildmat.2020.121942.

Anderson TR, Hawkins E, Jones PD. CO2, the greenhouse effect and global warming: from the pioneering work of Arrhenius and Callendar to today's Earth System Models. *Endeavour*. 2016;40(3):178-187. doi:10.1016/j.endeavour.2016.07.002.

Andrew RM. Global CO_2 emissions from cement production, 1928-2018. *Earth System Science Data*. 2019;11(4):1675-1710. doi:10.5194/essd-11-1675-2019.

Askar MK, Al-Kamaki YSS, Ferhadi R, Majeed HK. Cracks in concrete structures causes and treatments: A Review. *Mağallaẗ Ğāmi'aẗ Duhūk*. 2023;26(2):148-165. doi:10.26682/csjuod.2023.26.2.16.

Bisulandu BJRM, Huchet F. Rotary kiln process: An overview of physical mechanisms, models and applications. *Applied Thermal Engineering*. 2023;221:119637. doi:10.1016/j.applthermaleng.2022.119637.

Bojanovský J, Máša V, Hudák I, Skryja P, Hopjan J. Rotary Kiln, a unit on the border of the process and energy industry-current state and perspectives. *sustainability*. 2022;14(21):13903. doi:10.3390/su142113903. doi:10.3390/su142113903.

Cordoba G., Paulo C. I., Irassar E. F. Methodology for the evaluation of the environmental impact of elaborated concrete applied to the metropolitan region of Buenos Aires," Revista Hormigón, 2023, vol. 64.

de Rooij, M., Van Tittelboom, K., De Belie, N., & Schlangen, E. (Eds.). *Self-healing phenomena in cement-based materials: state-of-the-art report of RILEM technical committee 221-SHC: Self-healing phenomena in cement-based materials.* 2013. https://doi.org/10.1007/978-94-007-6624-2.

Fleming JR. Joseph Fourier, the 'greenhouse effect', and the quest for a universal theory of terrestrial temperatures. *Endeavour.* 1999;23(2):72-75. doi:10.1016/s0160-9327(99)01210-7.

Gartner EM, Sui T. Alternative cement clinkers. *Cement And Concrete Research.* 2018;114:27-39. doi:10.1016/j.cemconres.2017.02.002.

Gebru KA, Kidanemariam TG, Gebretinsae HK. Bio-cement production using microbially induced calcite precipitation (MICP) method: A review. *Chemical Engineering Science.* 2021;238:116610. doi:10.1016/j.ces.2021.116610.

Gentry MC, Jacobi AM. Heat transfer enhancement by delta-wing vortex generators on a flat plate: Vortex interactions with the boundary layer. *Experimental Thermal And Fluid Science.* 1997;14(3):231-242. doi:10.1016/s0894-1777(96)00067-2.

Griffiths S, Sovacool BK, Del Rio DF, et al. Decarbonizing the cement and concrete industry: A systematic review of socio-technical systems, technological innovations, and policy options. *Renewable & Sustainable Energy Reviews*. 2023;180:113291. doi:10.1016/j.rser.2023.113291.

Handore AV, Khandelwal SR, Karmakar R, Jagtap AS, Handore DV. Bioconcrete: The Promising Prospect for Green Construction. *Springer eBooks*. 2021; 567-584. doi:10.1007/978-3-030-76073-1_29.

Hernández-Santana A, Gómez-Garzón C, Dussán J. Complete Genome Sequence of *Lysinibacillus sphaericus* WHO Reference Strain 2362. *Genome Announcements*. 2016;4(3). doi:10.1128/genomea.00545-16.

IPCC, 2014: Climate Change 2014: Synthesis Report. Contribution of Working Groups I, II and III to the Fifth Assessment Report of the Intergovernmental Panel on Climate Change [Core Writing Team, R.K. Pachauri and L.A. Meyer (eds.)]. IPCC, Geneva, Switzerland, 151 pp.

IPCC, 2019: Summary for Policymakers. In: Climate Change and Land: an IPCC special report on climate change, desertification, land degradation, sustainable land management, food security, and greenhouse gas fluxes in terrestrial ecosystems [P.R. Shukla, J. Skea, E. Calvo Buendia, V. Masson-Delmotte, H.- O. Pörtner, D. C. Roberts, P. Zhai, R. Slade, S. Connors, R. van Diemen, M. Ferrat, E. Haughey, S. Luz, S. Neogi, M. Pathak, J. Petzold, J. Portugal Pereira, P. Vyas, E. Huntley, K. Kissick, M. Belkacemi, J. Malley, (eds.)]. https://doi.org/10.1017/9781009157988.001.

IPCC, 2021: Climate Change 2021: The Physical Science Basis. Contribution of Working Group I to the Sixth Assessment Report of the Intergovernmental Panel on Climate Change [Masson-Delmotte, V., P. Zhai, A. Pirani, S.L. Connors, C. Péan, S. Berger, N. Caud, Y. Chen, L. Goldfarb, M.I. Gomis, M. Huang, K. Leitzell, E. Lonnoy, J.B.R. Matthews, T.K. Maycock, T. Waterfield, O. Yelekçi, R. Yu, and B. Zhou (eds.)]. *Cambridge University Press*, Cambridge, United Kingdom and New York, NY, USA, 2391 pp. doi:10.1017/9781009157896.

Iqbal DM, Wong LS, Kong SY. Bio-Cementation in Construction Materials: A Review. *Materials*. 2021;14(9):2175. doi:10.3390/ma14092175.

Kahawalage AC, Melaeen MC, Tokheim LA. Opportunities and challenges of using SRF as an alternative fuel in the cement industry. *Cleaner Waste Systems*. 2023;4:100072. doi:10.1016/j.clwas.2022.100072.

Kayondo M, Combrinck R, Boshoff WP. State-of-the-art review on plastic cracking of concrete. *Construction & Building Materials*. 2019;225:886-899. doi:10.1016/j.conbuildmat.2019.07.197.

Laanaiya M, Bouibes A, Zaoui A. Understanding why Alite is responsible for the main mechanical characteristics in Portland cement. *Cement And Concrete Research*. 2019;126:105916. doi:10.1016/j.cemconres.2019.105916.

Li J., Zhang W., Li C. Montei P. J. M., Green concrete containing diatomaceous earth and limestone: workability, mechanical properties, and life-cycle assessment, *Journal of Cleaner Production*, 2019 223, pp. 662-679,

Madlool NA, Saidur R, Hossain M, Rahim NA. A critical review on energy use and savings in the cement industries. *Renewable & Sustainable Energy Reviews.* 2011;15(4):2042-2060. doi:10.1016/j.rser.2011.01.005.

Marcondes, Carlos Gustavo Nastari et al. *Lysinibacillus sphaericus* as a self-healing agent for cement-based materials: a preliminary investigation, *International Journal of Development Research.* 2021; 11 (02): 44195-44200. doi:10.37118/ijdr.21078.02.2021.

Meng Q, Zhu J, Wang T. Numerical Prediction of Long-Term Deformation for Prestressed Concrete Bridges under Random Heavy Traffic Loads. *Journal Of Bridge Engineering.* 2019;24(11). doi:10.1061/(asce)be.1943-5592.0001489.

Mohamad N, Muthusamy K, Embong R, Kusbiantoro A, Hashim MH. Environmental impact of cement production and Solutions: A review. *Materials Today: Proceedings.* 2022;48:741-746. doi:10.1016/j.matpr.2021.02.212.

Muller V, Pacheco F, Carvalho CM, et al. Analysis of cementitious matrices self-healing with bacillus bacteria. *IBRACON Journal of Structures and Materials.* 2022;15(4). doi:10.1590/s1983-41952022000400004.

Naveed M, Duan J, Uddin S, Suleman M, Hui Y, Li H. Application of microbially induced calcium carbonate precipitation with urea hydrolysis to improve the mechanical properties of soil. *Ecological Engineering.* 2020;153:105885. doi:10.1016/j.ecoleng.2020.105885.

Nusrat A, Shimizu M. *Lysinibacillus* Species: Their Potential as Effective Bioremediation, Biostimulant, and Biocontrol Agents. *Reviews In Agricultural Science.* 2021;9(0):103-116. doi:10.7831/ras.9.0_103.

Ortega-Villamagua E, Arcos M, Romero M, Mora CAV, Palma-Cando A. Microbiologically induced carbonate precipitation as potential strategy in the restoration of heritage structures. *Ge-conservation.* 2022;21(1):224-234. doi:10.37558/gec.v21i1.1119.

Paul VG, Wronkiewicz DJ, Mormile MR. Impact of elevated CO_2 concentrations on carbonate mineral precipitation ability of sulfate-reducing bacteria and implications for CO_2 sequestration. *Applied Geochemistry.* 2017;78:250-271. doi:10.1016/j.apgeochem.2017.01.010.

Pise N, Meshram T, Doijad Y, et al. A Brief Study on Causes of Cracks, Prevention and Pattern of Cracks on Concrete. *International Journal Of Scientific Research In Science, Engineering And Technology.* june 2021:439-443. doi:10.32628/ijsrset2183194.

Schuur EAG, McGuire AD, Schädel C, et al. Climate change and the permafrost carbon feedback. *Nature.* 2015;520(7546):171-179. doi:10.1038/nature14338.

Seifan, M., & Berenjian, A. (2018). Application of microbially induced calcium carbonate precipitation in designing bio self-healing concrete. *World Journal Of Microbiology And Biotechnology*, *34*(11). https://doi.org/10.1007/s11274-018-2552-2.

Soffritti I, D'Accolti M, Lanzoni L, et al. The Potential Use of Microorganisms as Restorative Agents: An Update. *Sustainability.* 2019;11(14):3853. doi:10.3390/su11143853.

Sohail MG, Wang B, Jain A, et al. Advancements in Concrete Mix Designs: High-Performance and Ultrahigh-Performance Concretes from 1970 to 2016. *Journal Of Materials In Civil Engineering*. 2018;30(3). doi:10.1061/(asce)mt.1943-5533.0002144.

Sun L, Pang X, Ghabezloo S, Wang H, Sun J. Hydration kinetics and strength retrogression mechanism of silica-cement systems in the temperature range of 110 °C-200 °C. *Cement And Concrete Research*. 2023;167:107120. doi:10.1016/j.cemconres.2023.107120.

Tayebani B, Said A, Memari AM. Less carbon producing sustainable concrete from environmental and performance perspectives: A review. *Construction & Building Materials*. 2023;404:133234. doi:10.1016/j.conbuildmat.2023.133234.

Triantafyllou GG, Rousakis T, Karabinis AI. Analytical assessment of the bearing capacity of RC beams with corroded steel bars beyond concrete cover cracking. *Composites Part B, Engineering*. 2017;119:132-140. doi:10.1016/j.compositesb.2017.03.036.

Van Belleghem, P. Van den Heede, K. Van Tittelboom, and N. De Belie, "Quantification of the Service Life Extension and Environmental Benefit of Chloride Exposed Self-Healing Concrete," *Materials,* vol. 10, no. 1, 2017.

UN Environment, Scrivener K. L., John V. M. Gartner E. M. Eco-efficient cements: Potential economically viable solutions for a low-CO2 cement-based materials industry, Cement and Concrete Research, 2018. 114, pp. 2-26

Wang, J. J. Wang Y. F., Sun Y. W. Tingley D. D. Life cycle sustainability assessment of fly ash concrete structures, *Renewable and Sustainable Energy Reviews*, 2017 80, pp. 1162-1174.

Wang R, Zhang Q, Li Y. Deterioration of concrete under the coupling effects of freeze-thaw cycles and other actions: A review. *Construction & Building Materials*. 2022;319:126045. doi:10.1016/j.conbuildmat.2021.126045.

Wei H, Fan Y, Du H, Liang R, Wang X. Regulation of Calcium Source and Addition Method for MICP in Repairing High-Temperature Concrete Damage. *Applied Sciences*. 2023;13(9):5528. doi:10.3390/app13095528.

Wen K, Yang L, Amini F, Li L. Impact of bacteria and urease concentration on precipitation kinetics and crystal morphology of calcium carbonate. *Acta Geotechnica*. 2019;15(1):17-27. doi:10.1007/s11440-019-00899-3.

Yoro KO, Daramola MO. CO2 emission sources, greenhouse gases, and the global warming effect. In: *Elsevier eBooks*. ; 2020:3-28. doi:10.1016/b978-0-12-819657-1.00001-3.

Zhang K, Tang CS, Jiang NJ, et al. Microbial-induced carbonate precipitation (MICP) technology: a review on the fundamentals and engineering applications. *Environmental Earth Sciences*. 2023;82(9). doi:10.1007/s12665-023-10899-y.

FSC
www.fsc.org
MIX
Papier | Fördert gute Waldnutzung
FSC® C083411

Kolibri 360 GmbH
Ferdinand-Jühlke-Straße 7
99095 Erfurt, Deutschland
produktsicherheit@kolibri360.de

Druck:
CPI Druckdienstleistungen GmbH
im Auftrag der
Kolibri 360 GmbH
Ferdinand-Jühlke-Str. 7
99095 Erfurt